RAINFOREST ANIMALS

MONKEYS

by Alissa Thielges

hand

tail

Look for these words and pictures as you read.

mouth

troop

What's swinging in the trees?
A monkey!

Many monkeys live in the rainforest. Some are big. Some are small.

hand

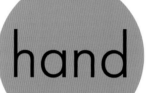

Look at the hand.
It has a thumb like you do.
It can grab fruit.

Look at the long tail.
It is like an extra arm.
It grips the branch.

tail

mouth

Look at the mouth.
A howler monkey is loud.
It can be heard miles away.

troop

Look at the troop.

It is a group of monkeys.

They live together.

A baby monkey is born.
It hangs onto mom.

hand

tail

Did you find?

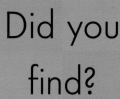

mouth

troop

Spot is published by Amicus Learning, an imprint of Amicus
P.O. Box 227, Mankato, MN 56002
www.amicuspublishing.us

Copyright © 2024 Amicus.
International copyright reserved in all countries.
No part of this book may be reproduced in any form without written permission from the publisher.

Library of Congress Cataloging-in-Publication Data
Names: Thielges, Alissa, 1995- author.
Title: Monkeys / by Alissa Thielges.
Description: Mankato : Amicus Learning, [2024] | Series: Spot rainforest animals | Audience: Ages 4-7 | Audience: Grades K-1 | Summary: "A search-and-find book about monkeys reinforces new vocabulary to build reading success while close-up images of animals in their natural rainforest habitat captivate young audiences. A great early STEM book to inspire learning about life science for kindergartners and first graders"—Provided by publisher.
Identifiers: LCCN 2023011696 (print) | LCCN 2023011697 (ebook) | ISBN 9781645492610 (library binding) | ISBN 9781681527857 (paperback) | ISBN 9781645493495 (pdf)
Subjects: LCSH: Monkeys--Juvenile literature.
Classification: LCC QL737.P9 T442 2024 (print) | LCC QL737.P9 (ebook) | DDC 599.8--dc23/eng/20230421
LC record available at https://lccn.loc.gov/2023011696
LC ebook record available at https://lccn.loc.gov/2023011697

Printed in China

Rebecca Glaser, editor
Deb Miner, series designer
Mary Herrmann, book designer
Omay Ayres, photo researcher

Photos by Dreamstime/Eric Gevaert, 14, Lukas Blazek, 12–13; Getty Images/Ger Bosma, 6–7, Justin Russo, 8–9; iStock/Anna-av, 1, Manakin, 3, Miroslav_1, 4–5, reisegraf, 10–11, Eric Isselee, cover, 16

MONKEYS